1 Compton Bay, Isle of Wight

What are fossils?

British fossils have attracted attention since before the dawn of history. Nearly a hundred fossil sea urchins from the Chalk were found surrounding two human skeletons in a Bronze Age barrow near Dunstable, Bedfordshire, and other sites of this age have yielded such fossils as belemnites, bivalves, sponges, crinoid stems and fish teeth. We will never know what Bronze Age Man thought of these objects, but the folklore current until recently in Britain may give us some idea. Among the poorer people fossils were used to cure the sick and to protect the healthy. So, for example, oysters of the genus *Gryphaea* (fig 2), which can look something like

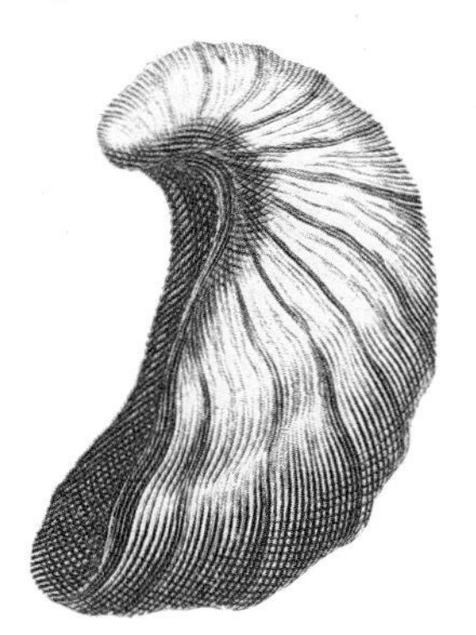

2 Stone like an oyster, 1677

an arthritic toe or finger, were used to cure joint pains; ammonites (fig 3), thought to be petrified snakes, gave protection against snakebite; sea urchins (fig 4) and belemnites, both thought to be thrown to Earth during thunderstorms, gave protection against lightning strike. Other fossils, such as the button-like teeth of the fossil fish *Lepidotes*, were worn as lucky charms.

During the 17th century, naturalists in Britain began to collect and study fossils and to speculate on their origins. The word 'fossil' was used in a broader sense than at present and included all sorts of shaped stones dug from the ground – concretions, crystals, implements and sedimentary structures as well as true fossils. John Woodward, a London physician, built up a collection of several thousand such fossils, which he described in a catalogue published in 1728. He bequeathed his collection to the University of Cambridge, where it is still preserved in the Sedgwick Museum of Geology. A fierce debate about the origin of fossils lasted from 1650 to 1750. Were these stones shaped like leaves, sea shells and bones formed by a force within the

3 Stone like a nautilus, 1677

Earth, or were they the remains of once-living animals and plants? If they had once been alive, how did the sea shells arrive at their present positions far inland and even on mountain tops? Naturalists whose local fossils were ancient, mineralised or present only as moulds, different from any known living forms and found far from the sea, tended to favour the first possibility. Figures 2–4 are from *The Natural History of Oxfordshire* (1677) by Robert Plot, who believed that fossils were 'naturally produced by some extraordinary plastic virtue latent in the Earth or quarries where they are found.' Robert Camden referred to fossils as 'little sporting miracles of nature' in 1586. Naturalists who studied less ancient fossils, such as those found at Barton, Hampshire,were much more inclined to accept the second possibility and to believe that Noah's flood had washed them into their present positions.

Interest in fossils declined in Britain during the 18th century, and, by the time it revived at the beginning of the 19th, the debate was over and naturalists accepted that fossils, by then distinguished from minerals and concretions, were the remains or traces of ancient animals and plants embedded in rock. A few people still believed that Noah's flood had distributed fossils over the world, but this idea too disappeared as more became known about the disposition of fossils in rock strata. The foundations of modern palaeontology – a word meaning the study of ancient life – were laid in the early 1800s with the work of naturalists such as James Sowerby (fig 5), James Parkinson and William Martin. Fossils were described, illustrated and given names in learned books published all over the world; they were studied in relation to the rock layers in which they were found and were used for dating and tracing them; later in the century fossils were used to build up a picture of ancient environments, to work out the course of evolution and to give clues to the interrelation of living things. Today there seems to be no end to the information that can be gleaned from these objects that were once carried around only for luck and protection.

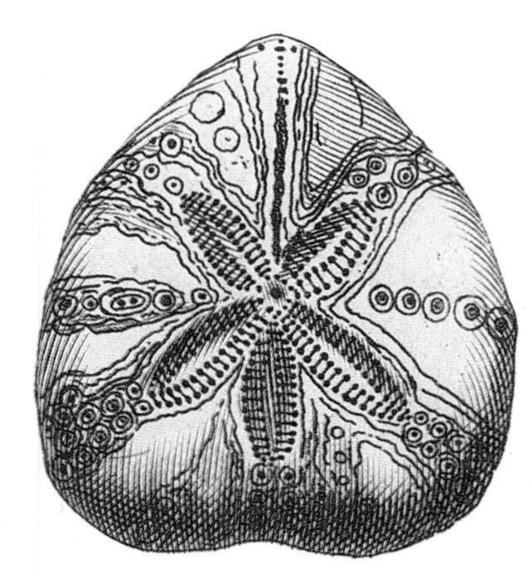

4 Stone like a sea urchin, 1677

5 Fossil from Essex, from *Mineral Conchology* by James Sowerby, 1818

How fossils form

Fossils are the remains and traces of ancient animals and plants preserved in rock. For a fossil to be formed, a living organism must either leave an impression of itself, shed its covering, or die in a place where sediment is accumulating, or else its remains must be carried into such a place after death. This sediment must then be buried by other layers and harden into rock deep underground. For the fossil to be found at the surface, the rock must be uplifted and exposed by erosion.

As most sediments are laid down under water, organisms living in lakes, river deltas or the sea stand a better chance of being fossilised than those in mountains or forests. And again, as soft-bodied animals and plants are almost always eaten or decay soon after they die, animals with shells or skeletons are more likely to be fossilised than those without. For these reasons, among others, the fossil record gives a very incomplete picture of ancient life.

Figure 6 shows some of the things that can happen to a shellfish after it has died. The soft parts of the animal, if not immediately eaten by scavengers, start to decay soon after death and, unless burial is rapid, the two parts of the shell will separate. Water currents may pick up the shell and carry it away, bumping and breaking it in the process (1). Once the shell is embedded, organic material starts to decay within its structure, making it lighter and weaker than when the animal was alive. Percolating water in a sand or other permeable sediment may dissolve the shell (2); this is one reason why sandstones only rarely contain fossils. Bodies of sediment are frequently eroded away as water currents change, particularly in rivers, small lakes and in very shallow sea water. Such erosion may either destroy the shell or else deposit it far from the place where it lived (3).

As the sediment hardens, tiny spaces within the shell become filled with a mineral such as calcite from water in the rock. This process, 'permineralisation', makes the shell heavier and stronger again (4). Any large spaces free of sediment may be filled at this stage. Other elements in the percolating water may come out of solution to replace the calcium carbonate of the shell, giving rise to fossils made of such minerals as chalcedony, opal, hematite, cassiterite and pyrite (5). Alternatively, the shell may be dissolved by water in the rock, leaving a natural mould which preserves the shape of the shell (6); if the empty mould is later filled by a new mineral, the result will be a natural cast of the shell (7). Under certain conditions the shell will recrystallise so that the original complex structure becomes a mass of large, interlocking crystals (8).

Muddy sediments are usually compressed as they harden into rock, and the shells within them are generally crushed (9), although they may retain their shape within a hard calcareous nodule. If the rock is deformed or strongly heated, the fossil will be distorted or may be completely obliterated (10).

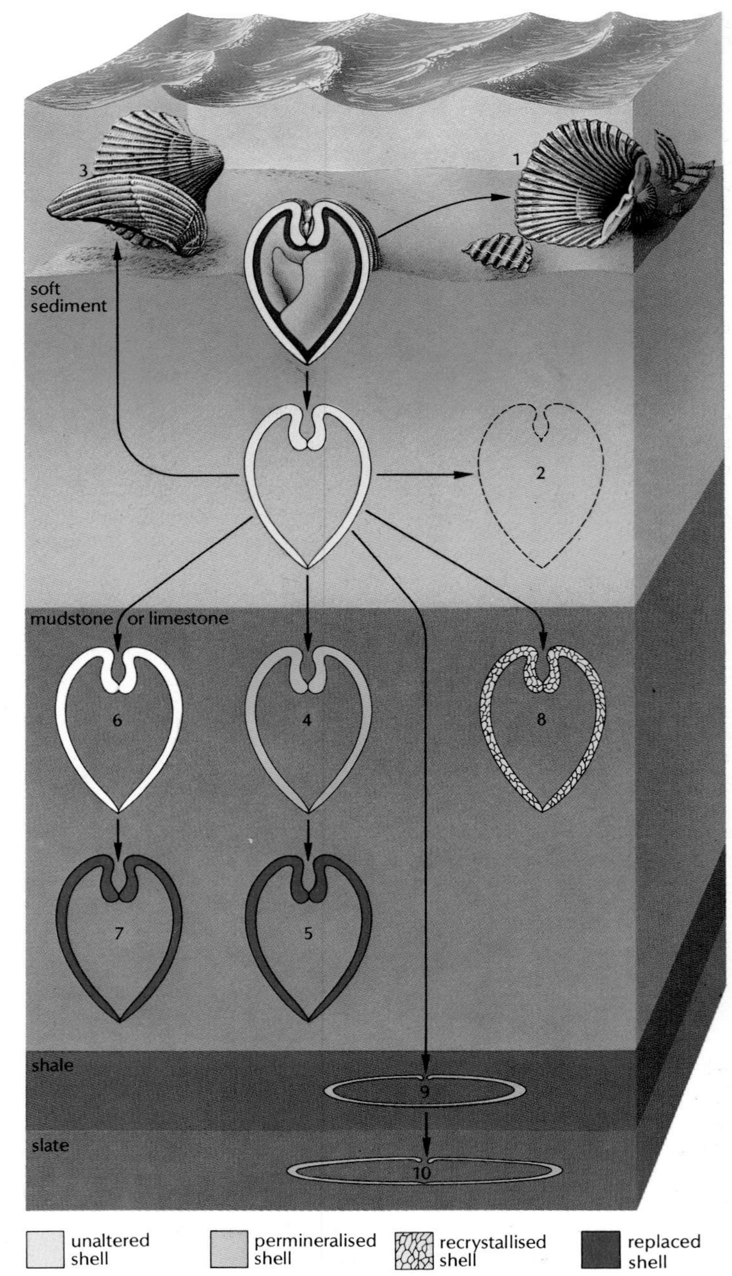

6 Some ways in which fossils are formed

Classification

Life on Earth is not a smooth continuum but is made up of distinct plant and animal types which are distinguished from one another by differences in shape or colour. Thus, for example, red deer, roe deer and fallow deer are distinct types; intermediate forms do not exist because the deer have evolved to fill different ecological niches. Red deer are not all identical, but, because they interbreed, there is a complete gradation between varieties. Such interbreeding animal groups, the natural units of classification, are called 'species' and are defined on anatomical, behavioural and geographical grounds. Species of animals and plants which closely resemble each other are grouped into 'genera', and these successively into 'families', 'orders', 'classes', 'phyla' and 'kingdoms'. Each of these categories is given a Latin name, but a particular organism is referred to by only its generic and specific names along with the name of the scientist who first described it: the red deer is *Cervus elephas* Linnaeus. Naming is governed by a complex set of rules, and each new species must be named and described in a particular way, with a 'type specimen' to act as the official name-bearer.

Palaeontologists attempt to fit fossils into the same grand scheme. They have very little behavioural data, and the anatomical and geographical information is always incomplete. Many fossil species were first named on the basis of a small part of the original animal (fig 8), while others are named even though the nature of the original animal or plant is unknown (fig 9). Palaeontologists do, however, have the advantage that they can study species through time and can sometimes trace their evolution. They attempt to reconstruct the overall shape of the 'tree of life', while biologists look in detail at just the tips of the branches – the living things of the present day. Figure 7 shows part of the evolutionary tree of the genus *Argopecten* and its subdivision into twelve species, five of which are alive today.

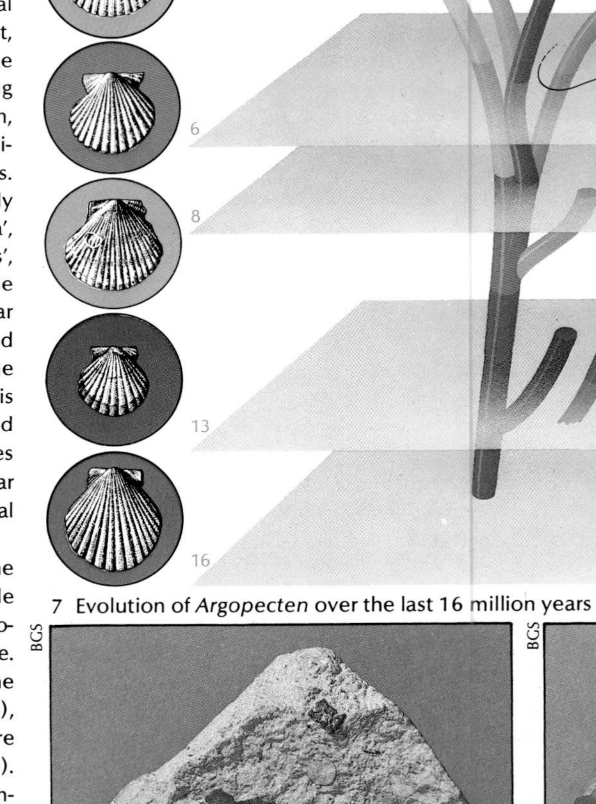

7 Evolution of *Argopecten* over the last 16 million years

8 *Triconodon*, known at first only by jawbone

9 *Palaeoxyris*, a plant or a fish egg-case?

Fossil families

Plants

Plants have an immensely long history. The oldest fossils, found in Australia and southern Africa, are minute bacteria and primitive single-celled algae preserved in chert (bedded flint) nearly 3500 million years ago. Bun-shaped masses of banded rock called 'stromatolites', which were formed by algae, are known from rocks 2700 million years old. Similar structures are forming today, in the hot, salty waters of Shark Bay, Australia, for example. More advanced algae with a cell nucleus and other features are first found in rocks about 600 million years old.

The oldest-known British fossils are tiny algal spores called acritarchs, which are found near the base of the Torridon Sandstone in western Scotland, and which are about 1000 million years old. Stromatolites 600–700 million years old have been found in the Dalradian of Scotland and in Anglesey. Marine algae are present in Lower Palaeozoic rocks, although freshwater forms are not found below the Devonian System.

Tiny plants, about the size of a pin and reproducing by spores, colonised the land early in the Silurian Period, and larger 'vascular' forms appeared soon after. These are particularly well preserved in beds of early Devonian chert near Rhynie, Scotland, which represents an ancient delta marsh. The three main types, loosely grouped as 'psilopsids', were all small plants with slender branching stems which carried spore cases at the tips or along the sides. The stems had internal tubes for carrying water from the earth throughout the plant – a vascular system – and a tough surface cuticle with openings for gas exchange known as stomata. The plants had no true leaves or roots and many must have looked very like one of the few living members of the group, *Psilotum*, the king fern.

Three groups of plants evolved from the psilopsids during the Devonian Period. They developed leaves and roots and the ability to thicken their stems with wood and so grow into shrubs and trees. Like the psilopsids, they carried spores which, on ripening, developed first into tiny plants quite unlike the parent. These reproduced by sex cells called gametes. If the plant surface was wet enough, the male cells swam across to fertilise the female cells, one of which eventually developed into a large spore-bearing plant. This 'alternation of generations' is seen in the living descendants of these three groups, the ferns (fig 10), club mosses and horsetails. Living members of these groups are mostly small, but in the past, particularly during late Carboniferous times, they included large trees in forests which covered much of the land and gave rise to our coal deposits.

COLEMAN

10 Tree ferns growing in New Zealand

COLEMAN

11 Cycads growing in Australia

COLEMAN

12 Musk mallow flower

These spore-bearing plants, with their need for a wet place for the gamete-bearing generation, represent an incomplete adaptation to life on land, in some ways analogous to the amphibians among the vertebrates. In a fourth, more completely adapted group, which evolved during the Devonian, airborne male cells fertilised a large female structure (the ovule) carried on the plant. When fertilised, this dropped to the ground as a seed and grew into a new plant. This group is known as the 'gymnosperms' because the female cell was carried exposed (*gymnus* – naked) on a cone. Members of the group are extinct cordaites and seed ferns, both common in the Carboniferous forests; cycads (fig 11), ginkgos and bennetitales, which were important in the Mesozoic and are still represented by a few descendants; and the still-flourishing conifers.

Much later a further evolutionary step was taken in which the ovule became enclosed in an ovary which was displayed, together with the male cells, in a 'flower', a structure developed to facilitate fertilisation of one plant by another with the aid of the wind, insects or birds. Flowering plants – 'angiosperms' – dominate the landscape today and have done so for the last 50 million years (fig 12). The male cells, 'pollen-grains', are resistant to decay and are preserved in sediments that are not otherwise fossiliferous. They are much more diagnostic of particular plants than leaves or twigs, so their study, 'palynology', is of great importance, particularly in the search for oil and gas (p 12).

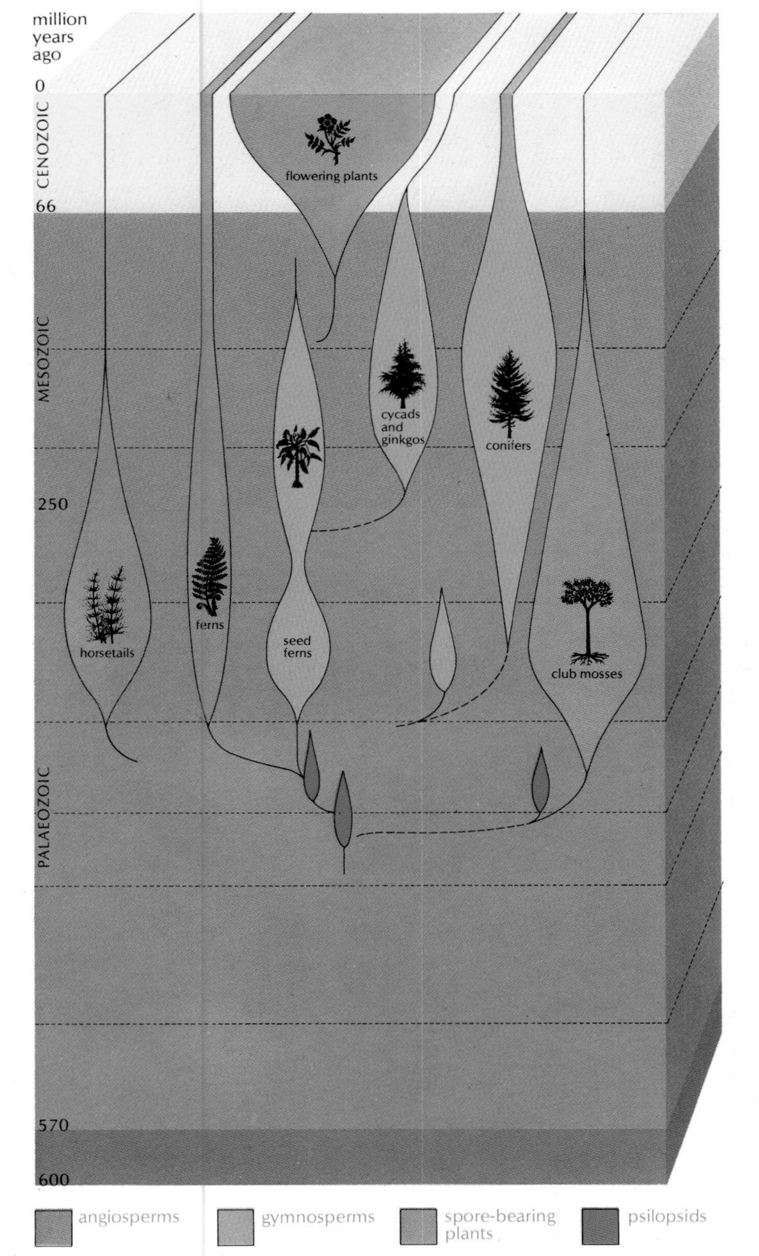

13 Evolution of vascular plants

BM(NH)/BGS

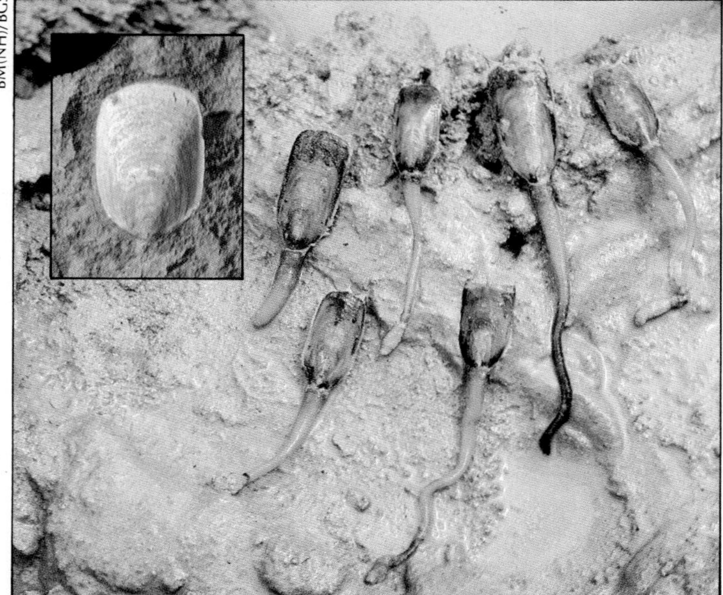

14 Living and (inset) fossil brachiopods, both *Lingula*

COLEMAN

15 Horseshoe crab, closest living relative of the trilobites

Animals without backbones

The oldest traces of animals are found in rocks about 600 million years old. They are the impressions of soft-bodied creatures, including jellyfish and their relatives, segmented worms, a few arthropods, and some other forms unlike anything alive today. About 570 million years ago, at the beginning of the Cambrian Period, one after another of these groups developed shells or skeletons, and within 50 million years all except one of the shelly phyla we know today had acquired hard parts.

Brachiopods were one of the first. The earliest were small, but by the Silurian Period the group was very important and included many different species. It has declined in numbers and variety since then, but there are still 300 species in the sea today, some of them very similar to fossil forms (fig 14). A brachiopod starts life as a free-swimming larva, settles on the sea floor after a few days and develops a shell made up of two valves, the dorsal (back) usually being smaller than the ventral (front). Most attach themselves to hard objects on the sea floor by a short muscular stalk, but others burrow, lie loose on the sea floor, or anchor themselves securely with the help of long spines. After the animal's death the valves often remain joined, though spines and delicate frills of shell generally break off.

Also flourishing in early Palaeozoic seas were the trilobites, an extinct class of the arthropod phylum related to the living horseshoe crab (fig 15). They first appeared early in the Cambrian, reached a maximum in the Ordovician and steadily declined in numbers until their extinction in the Permian Period. Trilobites were segmented animals with a flexible body and rigid head and tail-shield. Their close-fitting jointed shell was thick on the back but very thin around the limbs, which are only very rarely preserved. The shell did not grow with the animal but was regularly shed and a larger one developed. Most fossil specimens are parts that have been shed rather than the complete shell of a dead animal. Trilobites were very varied in their form: some were smooth and some spiny; some had large compound eyes and some were blind; some were 65 cm long, some were hardly more than 1 mm. Most trilobites walked or swam close to the sea floor in shallow water, though some burrowed and some floated. Long tracks have been found that are thought to have been made by walking trilobites.

A third group that appeared later in the Cambrian, just over 500 million years ago, is the cephalopods, one class of the mollusc phylum. Earliest forms had a squid-like body in a slightly curved, conical shell. The animal's head protruded through the open end of the cone, and the pointed end was divided internally by flat walls into buoyant gas chambers. These animals, whose only living descendant is the *Nautilus* of the Indian Ocean (fig 16), are called 'nautiloids'. From the nautiloids evolved animals with coiled shells and crinkly internal walls. Best known of these are the ammonites, whose beautifully ornamented shells are found in many

British Mesozoic rocks. An ammonite hatched as a larva which floated near the surface of the sea. A tiny shell developed, which coiled as the animal grew and which had a body chamber at the open end and partitioned gas chambers towards the centre. At intervals, as coiling continued, the animal moved forward in the body chamber and a new internal wall was secreted to form an extra gas chamber. Growth seems to have slowed down and stopped after about 20 years in most species. Adult males were much smaller than the females, and often developed shelly flaps called lappets at the aperture of the shell. Ammonites could move up and down in the water by changing the volume of gas in their chambers, but they were probably sluggish swimmers. For this reason it is thought they fed on plankton and larger plants, though some were provided with strong jaws and were probably also carnivorous or scavenging. Traces of gills, egg sacs and ink sacs have been found, but the number and form of the tentacles is unknown. Another group of cephalopods, which evolved from the nautiloids nearly 300 million years ago, includes the living squid and octopus and the extinct belemnite, known to fossil collectors by its bullet-shaped internal shell.

Among other invertebrate groups that are well known today and are common as fossils are the sea urchins (echinoids), sea lilies (crinoids), snails (gastropods), bivalves, corals and sponges.

16 Living nautiloid, *Nautilus*

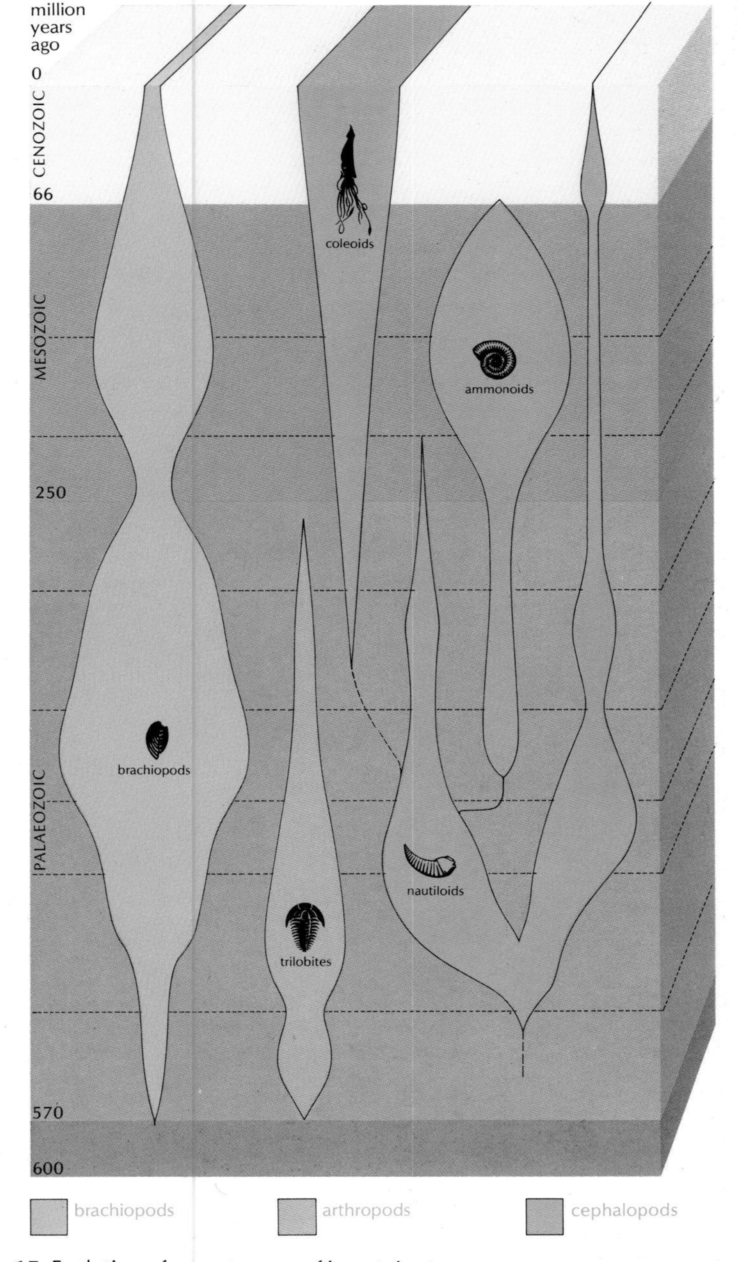

17 Evolution of some groups of invertebrates

Vertebrates and their relatives

Cats and dogs, birds, snakes, frogs and fish are all vertebrates. They all have an internal skeleton, a brain enclosed in a skull and a dorsal central nervous system. In addition there are certain small marine animals, including acorn worms and *Rhabdopleura*, which are clearly not vertebrates but which do possess features in common with them and are thought to be related. The extinct graptolites have hard parts rather like those of *Rhabdopleura*, and are therefore placed in the same group.

Graptolites were colonial animals that lived in the sea and secreted an external skeleton of protein. The graptolite started life as a free-swimming larva that developed a conical skeleton. A bud grew from the side of the cone and formed a cup in which lived a little animal (zooid). A second cup developed from a bud at the base of the first, and so on to form a branching colony up to 40cm across containing hundreds of zooids. The form and habits of the zooids are unknown, but using *Rhabdopleura* as a guide, figure 18 shows them with prominent food and oxygen-gathering arms, laying down layers of protein to strengthen the skeleton. There are two main groups of graptolites: the net- or bush-like 'dendroids', most of which lived attached to the sea floor, and the smaller 'graptoloids', with a simpler and more regular shape which floated near the surface of the sea.

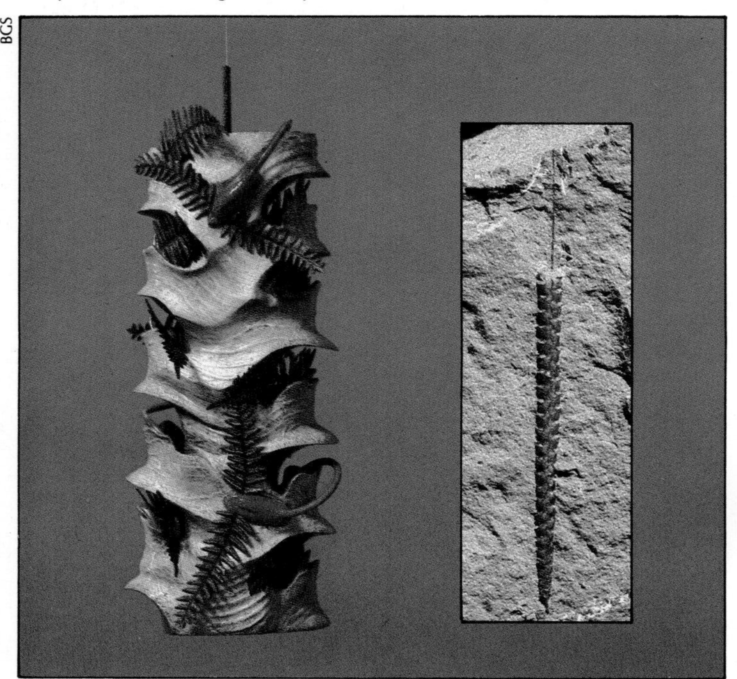

18 Model and (inset) fossil Ordovician graptolite, *Climacograptus*

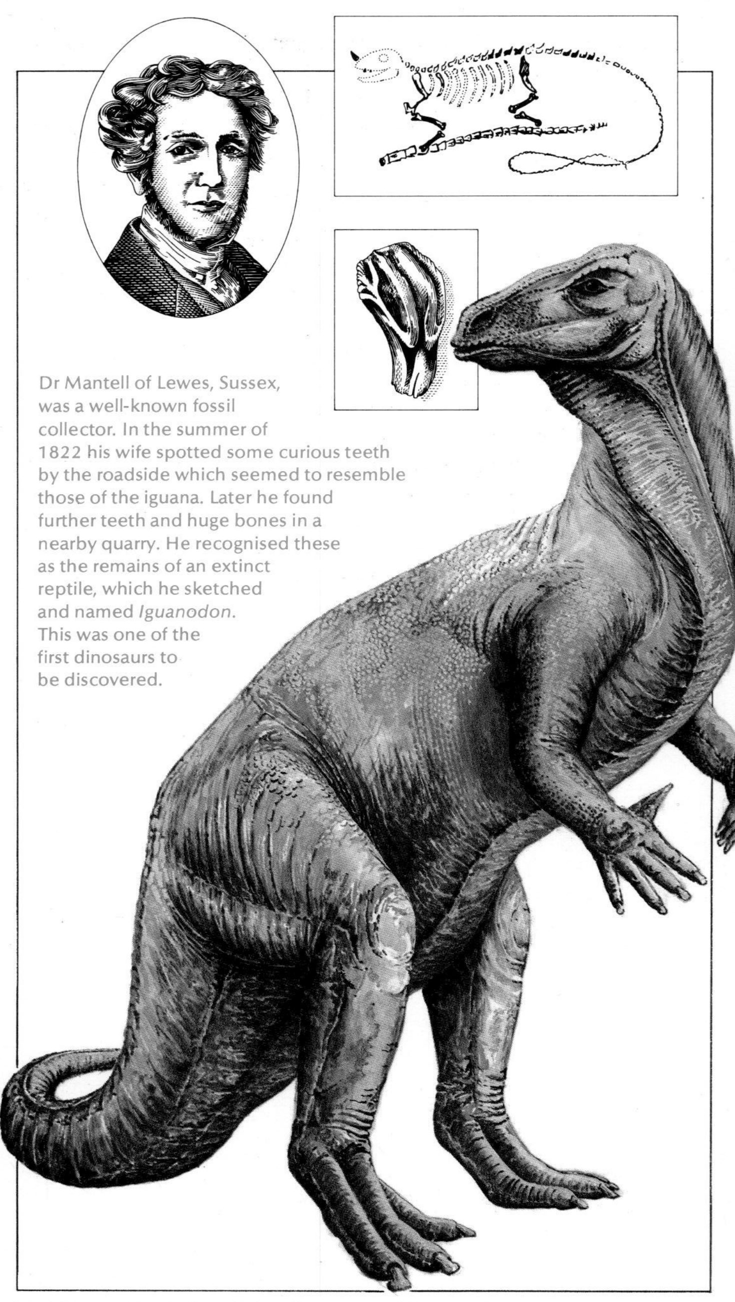

19 Gideon Mantell and the discovery of *Iguanodon*

The earliest-known vertebrates are the 'agnathans', fish that had no jaws or teeth but sucked small food particles up through a circular mouth. They were small animals with an internal skeleton of cartilage, an armour of bony plates or scales, and poorly developed fins. The group includes freshwater, brackish and marine forms, and is represented at the present day by the hagfish and lamprey (fig 20). Agnathan scale-fragments are found in rocks 510 million years old in Wyoming, USA; the oldest British specimens come from the middle Silurian rocks of Scotland and are about 420 million years old. Fish with jaws and teeth, paired fins and a bony internal skeleton first appeared at the end of the Silurian Period and rapidly evolved into a great variety of forms.

The subsequent evolution of the vertebrates includes at least three crucial phases. First was the development of lungs and five-fingered limbs by the amphibians, animals that live on land but have to return to water to breed and for their young to develop; second was the evolution of a waterproof egg by the reptiles, animals that can therefore spend all their time on land (fig 19). Third was the increase in size and complexity of the brain which, along with the evolution of the placenta and a constant body temperature, has characterised the placental mammals since they first appeared 80 million years ago. This trend is particularly marked in the primates and reaches its greatest development so far in Man.

COLEMAN

20 Living agnathan, a lamprey

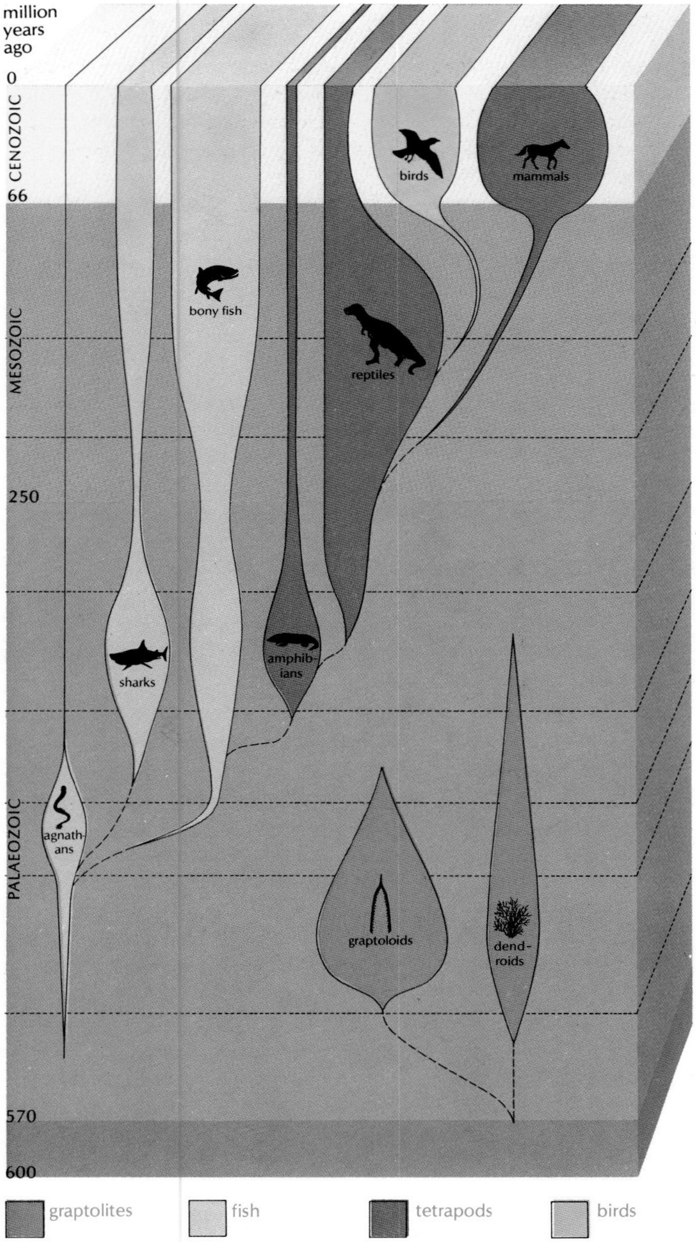

million years ago
0
CENOZOIC
66
MESOZOIC
250
PALAEOZOIC
570
600
birds
mammals
bony fish
reptiles
sharks
amphib-ians
agnath-ans
graptoloids
dend-roids
graptolites
fish
tetrapods
birds

21 Evolution of the vertebrates and graptolites

Scientific value of fossils

Fossils are interesting as a record of ancient life and many are aesthetically pleasing. To the geologist they can give important information about the age of the rocks in which they are found, about the environment in which those rocks were formed, and about broad features of the geography of the globe.

Different animals and plants have existed at different times in the Earth's history, and so their fossils can be used to indicate the geological age of the rocks in which they are found. On a broad scale trilobites existed from early Cambrian to late Permian times (fig 17) and so any rock in which they are found must belong within these age limits. Ammonites, belemnites and graptolites can be used in the same way to give even an inexperienced fossil collector some clue to the age of the rocks he is studying. On a much finer scale species of these various groups are used by palaeontologists to indicate the age of rock layers. Thus the Jurassic rocks of Britain are currently divided into 74 zones, each zone being named after an ammonite species. Whenever the ammonites defining a zone are found in a rock stratum, it can be assigned to that zone and its age, relative to other zoned strata in the Jurassic, is therefore known (fig 22). This is all part of the large and complex exercise of comparing strata in different parts of the country in terms of their rock-type and age, which is known as correlation. It is further explained in the companion booklet *The age of the Earth*. Correlation using fossils to establish age is called biostratigraphy.

Among the fossils most suitable for biostratigraphy are graptolites, trilobites, ammonites and their more ancient relatives the goniatites. These all evolved sufficiently rapidly for each species to exist for only a relatively short period; they floated or swam and so were not restricted to one particular environment and therefore are not found only in one rock type; and they are reasonably common and not too difficult to recognise. In their absence, however, many other types of fossils have been used, including fish, bivalves, sea urchins, brachiopods and belemnites.

Many rocks do not contain such fossils and the rock chips recovered from boreholes drilled in search of oil and gas are too small to contain recognisable fragments. Yet a detailed knowledge of the underground rock layers and their ages is essential if these fuels are to be found. The use of microscopic fossils has therefore become of great importance. The remains of many different microscopic animals and plants are found in sedimentary rocks. Tiny arthropods (ostracods, fig 23a), teeth-like structures from unknown animals (conodonts, fig 23b), single-celled animals (foraminifera, fig 23c), single-celled plants (diatoms, dinoflagellates, fig 23d, acritarchs and coccoliths), and plant spores and pollen are all used. Some of these are just as satisfactory as large fossils for use in biostratigraphy; others do not seem to be as precise, but may become more useful as knowledge of them increases.

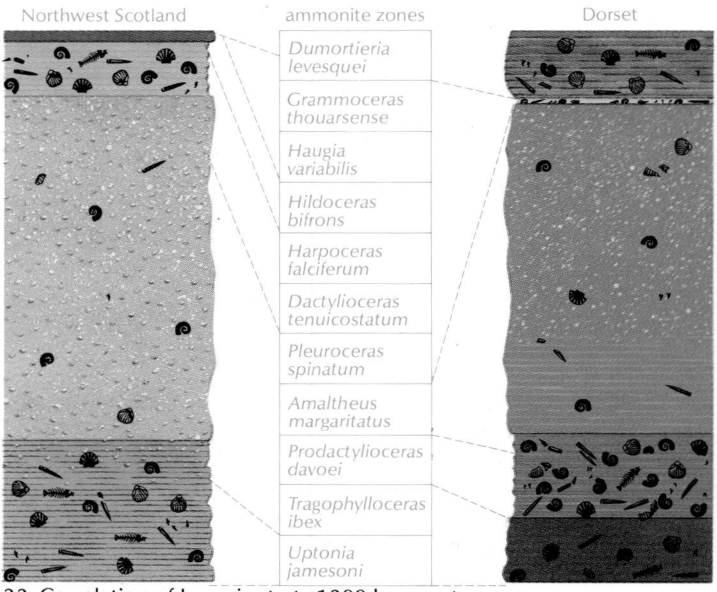

22 Correlation of Jurassic strata 1200 km apart

BCS

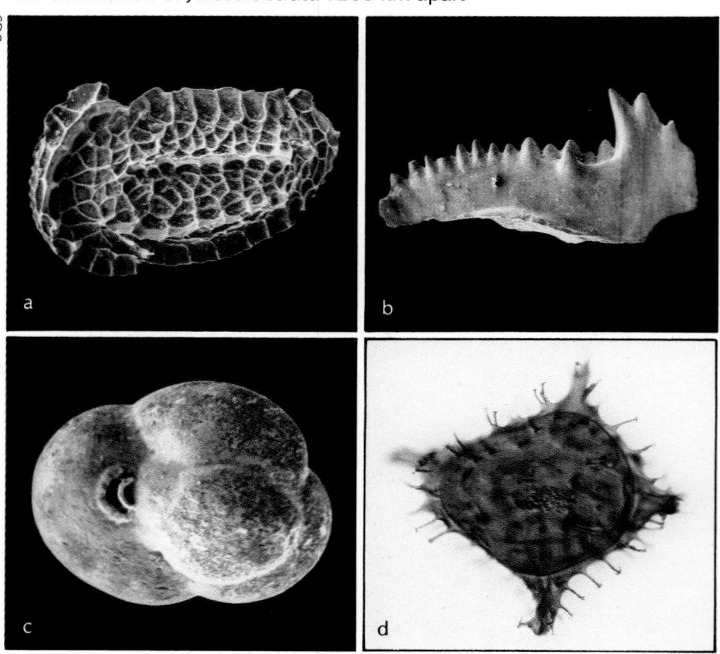

23 Microfossils used in correlation (all much enlarged)

Ancient environments. Fossils can provide information about ancient environments if the lifestyles of their closest living relatives are studied. So, since all living echinoids are marine, it may be assumed that strata containing fossil echinoids were laid down in the sea. Figure 24 shows hexacorals from a Jurassic coral reef at Steeple Ashton, Wiltshire. Since living hexacorals only form reefs in warm sea water less than 70m deep, similar conditions may well have existed around Steeple Ashton 160 million years ago. Deductions like this are less secure when fossil and living animals are more distantly related. The Silurian reefs at Dudley were built up partly by tabulate and rugose corals, which are now extinct, and which might have favoured different conditions from hexacorals. The brachiopod *Lingula*, on the other hand, has changed little since the early Palaeozoic and seems to have tolerated brackish water throughout the last 450 million years (fig 14).

Evolution. The succession of fossils in the strata that make up the geological column gives a picture, albeit incomplete, of the evolution of life over the last 570 million years, and a sketch of what might have happened earlier. Fossils are important, but not the only evidence of evolution. However, when fossils are collected from a succession of beds at a locality, it is unusual to find any signs of evolution in action. Changes in the fauna and flora are usually due to new forms migrating into an area or old ones departing. This is because evolution seems to occur in sudden bursts which affect small and isolated populations: gradual change over a long span of time may be the exception rather than the rule. Figure 25 shows ammonites from three levels within the *Asteroceras obtusum* Zone of the Lower Lias that are thought to be members of an evolving lineage. The zone consists of 12m of strata and the lineage represents a time span of less than one million years.

Continental drift. The distribution of fossil species gives many clues to the disposition of land and sea in the past and is one of the lines of evidence for continental drift. Early Ordovician fossils from Durness in Scotland (fig 26, top) are quite different from those found in rocks of the same age and general environment at Shelve, Shropshire (fig 26, bottom), and elsewhere in England. They are, however, similar to those from the eastern United States and northwest Ireland. This supports other evidence which indicates that Britain was cut in two by a wide ocean in the early Palaeozoic, and that Scotland and part of Ireland were joined to North America. Similarly, there are links between the Permian and Triassic fossils of South America, Africa, Madagascar, India, Antarctica and Australia, continents we believe were joined together just over 200 million years ago. Leaves of the tree *Glossopteris* have been found in all six places, and bones of the reptile *Lystrosaurus* in three.

BGS

24 Jurassic corals, Steeple Ashton, Wiltshire

BGS

25 Evolutionary series of Jurassic ammonites

BGS

26 Ordovician fossils, Scotland and England

Collecting fossils

The increasing popularity of geology in schools, colleges and among the general public has brought with it many problems. Many of the new enthusiasts want to build up their own collections of rocks, minerals and particularly fossils. Well-known localities are visited by thousands each year and attacked with hammers, chisels and even pneumatic drills until nothing of interest is left, or until the landowner, in desperation, closes the site. This is making both serious research and responsible amateur fossil collecting increasingly difficult.

Many people are becoming aware of the need for geological conservation. It is now realised that Britain's wealth of fossiliferous rocks must be both protected and developed. On the one hand, a code for geological fieldwork, prepared by the Geologists' Association and summarised in figure 28, aims to minimise the damage to sites and their surroundings and to ensure that site owners are not antagonised. Copies are available from the Geological Museum on request. On the other hand, efforts are being made by bodies such as the Nature Conservancy Council to increase the number of sites available. They have cleared some that are overgrown or otherwise difficult of access (fig 27), and have publicised others that are little known to collectors, in order to reduce the pressure on the famous ones. In addition a critical eye is kept on all planning applications for work such as the filling of quarries or the building of sea defences that would involve the loss of geological sites, and objections lodged where necessary.

NATURE CONSERVANCY COUNCIL

27 Clearing an overgrown quarry, Chilmark, Wiltshire

R GEARY

1 *Obey the Country Code and observe local byelaws. Remember to shut gates and leave no litter.*

2 *Always seek permission before entering private land.*

3 *Don't litter fields or roads with rock fragments which might cause injury to livestock or be a hazard to pedestrians or vehicles.*

4 *Don't take risks on or below insecure cliffs or rock faces.*

5 *Do not hammer to no purpose.*

6 *Keep collecting to a minimum. Collect from scree, fallen blocks or waste tips rather than from the actual rock face. Leave things for the next visitor to see.*

7 *Never collect from walls or buildings. Take care not to undermine fences, walls, bridges or other structures.*

8 *In quarries the wearing of safety helmets is recommended and may be insisted upon by the managers.*

9 *Take proper care of your specimens once you have collected them.*

28 A code for geological fieldwork

The importance of conservation does not mean that all fossil collecting is to be discouraged. If pursued wisely, causing no unnecessary damage, it can bring satisfaction and knowledge to the collector, and may even lead to important discoveries. The first steps are to get hold of geological maps and guide books covering the area to be visited (fig 29a), to go to a local museum to see the sorts of fossils that may be found, and to gather the necessary equipment (see below). Once out at a rock exposure, only experience and a good eye are needed to spot the fossils (fig 29b).

Hammer and chisel are used to trim down large fossil-bearing blocks, but no attempt should be made to extract the fossils at this stage. Specimens are wrapped in layers of tissue and newspaper, and the precise locality written on the paper. It is useful to make notes, draw sketches and take photographs of the places where the specimens are collected, as this will add to the interest and value of the collection. Once back home, the surface of a fossil can be carefully cleared of rock, though it is a good thing to leave some on the specimen (fig 29c). Fossils can be stored in cardboard or plastic trays, padded where necessary, with a label bearing at least the precise locality. Other information for the label includes the name and age of the rock formation and the Latin name of the fossil – books, museums and knowledgeable friends will all be helpful here (fig 29d).

29 Four stages of a fossil-collecting expedition

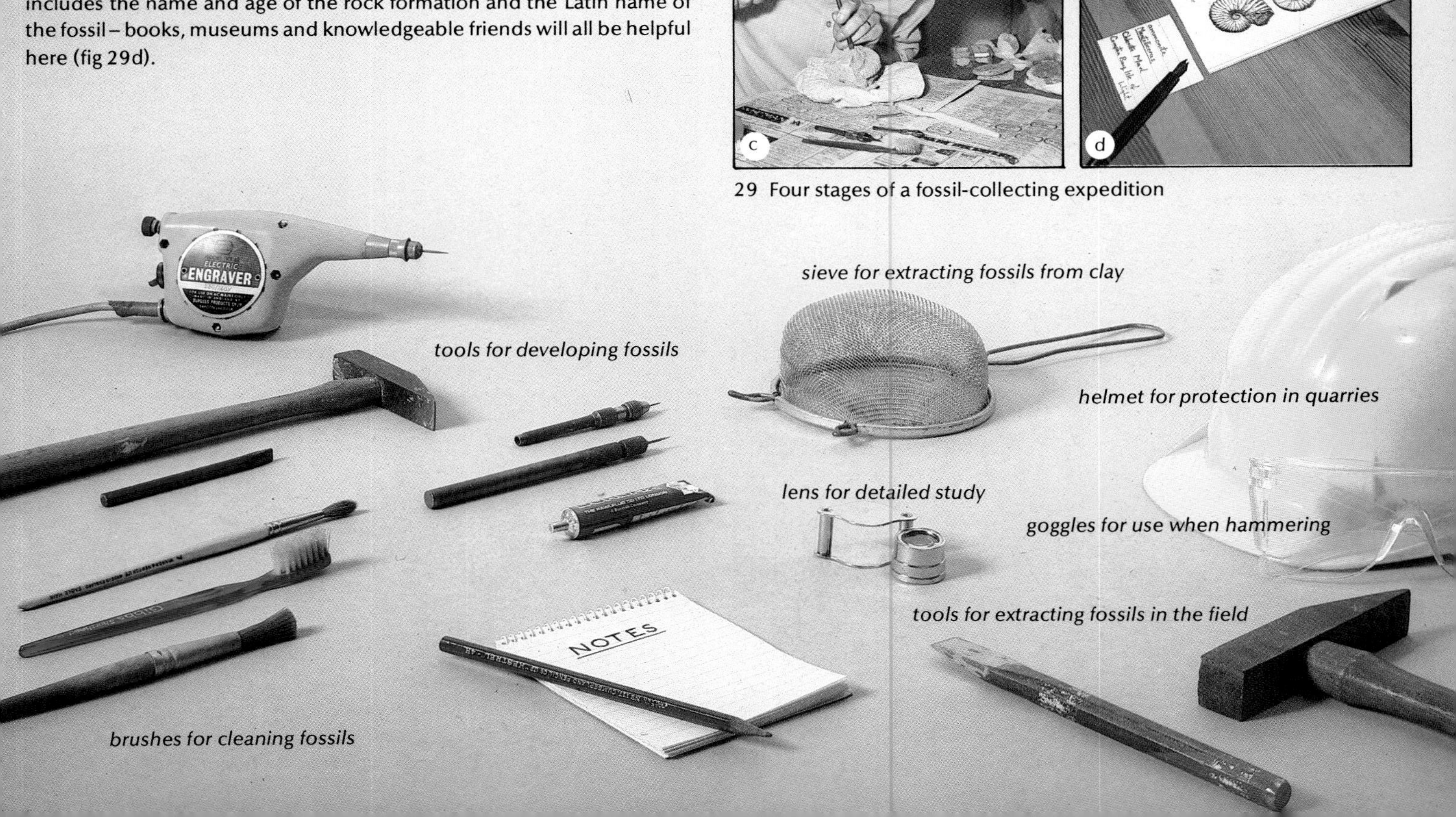

Fossils in museums

For anyone with a real interest in fossils, reading and collecting should be backed up by visits to museums. More than 250 museums in Britain have collections of fossils. The largest and most nearly comprehensive are the Natural History Museum and the Geological Museum, both in London. The Natural History Museum has a vast collection of fossils of all sorts and all ages from all over the world, and a large staff of palaeontologists. Mounted dinosaur skeletons stand in the Main Hall, and smaller fossils have a place in a number of the museum's new exhibitions dealing with the diversity and interactions of living things. The Geological Museum, at present part of the British Geological Survey, has a large display of British fossils arranged by age (fig 31) and a series of regional exhibits. The Survey has a huge collection of British fossils of all ages and sorts, and a staff of palaeontologists who identify fossils for the Survey's field geologists, and who apply their knowledge to problems connected with the search for fuels and other useful raw materials. Next in size and scope come the museums in the three regional capitals, Belfast, Cardiff and Edinburgh, and in the largest of the university and city museums, in Cambridge, Glasgow, Manchester, Oxford and York. However even small museums, such as those in Bath or Whitby, Ludlow or Sandown, may have collections of international importance and displays of more than local interest.

Museum collections have been built up over the last 200 years by the purchase or gift of specimens, either singly or by the thousand, from amateur fossil collectors and palaeontologists, and by the work of the museums' own staff. They may include type specimens (p.5), figured specimens (those illustrated in a book or paper), and fossils which are the basis of a geological map or of some piece of palaeontological research. There may be specimens from localities that are no longer accessible or that are 'worked out', and there may be rare or unnamed species.

It is the duty of the curator to add to these collections when he can, to keep them and the information associated with them safe for the future, to make specimens available for research, and to use them for the education of the general public by exhibitions, demonstrations and by assisting with identifications (fig 30). The needs of preservation, research and education often conflict and must be carefully balanced.

Museums are more than just exhibition halls, and certainly not just stores of old and unwanted specimens. They contain a rich and irreplaceable heritage, which is the raw material for our present understanding of ancient life and of biostratigraphy, and which will form the basis for a great deal of future research. Museums are also a vital part of geological education, giving people a chance to see fossils they could never hope to find, and perhaps sparking off an interest in the Earth and its history.

BGS

30 Fossil identification at the Geological Museum, London

BGS

31 British Fossils exhibition, Geological Museum, London

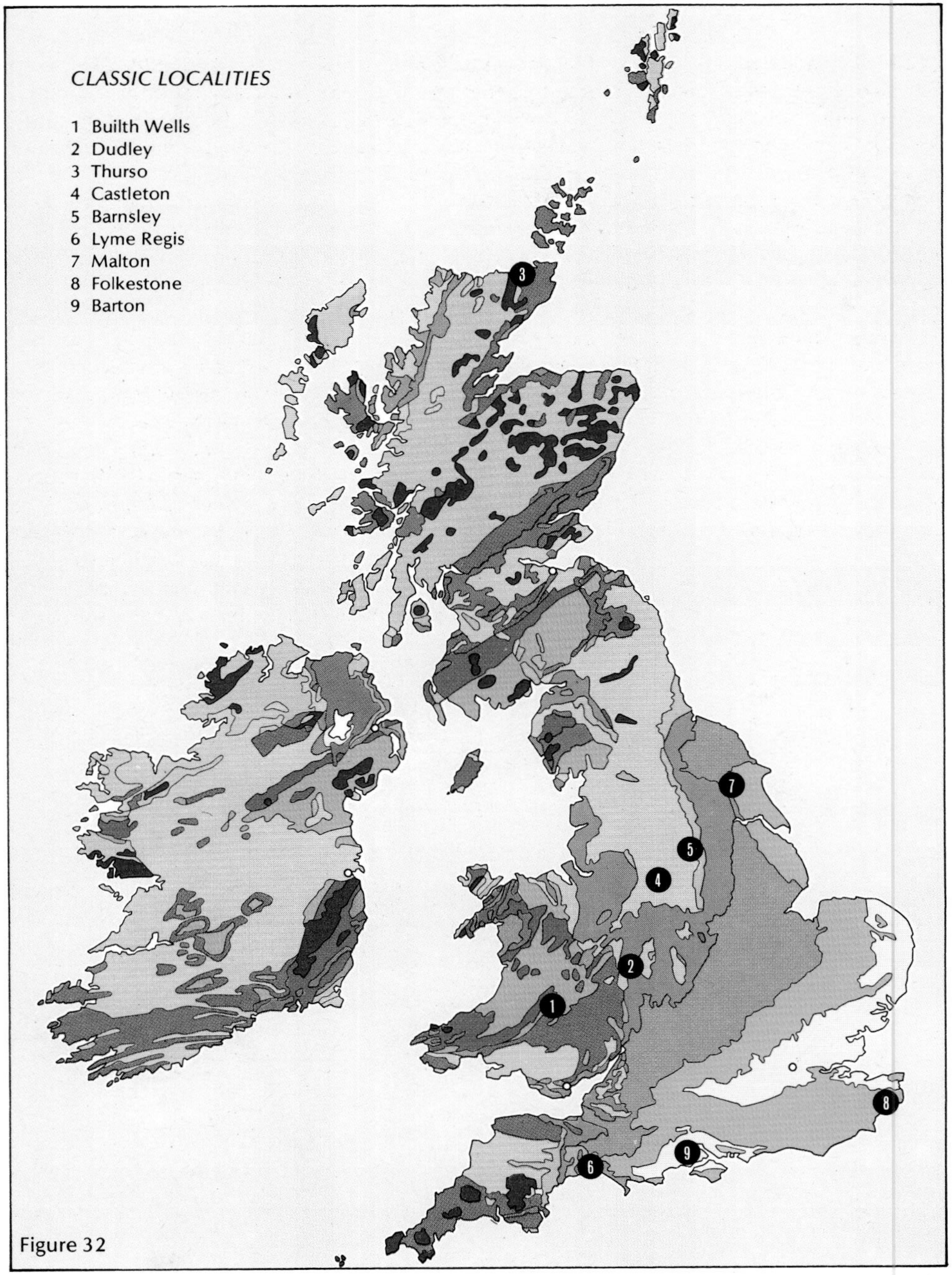

Classic localities

There are many thousand sites in Britain where fossils have been found. Some are large permanent exposures such as sea cliffs or mountain crags, some are quarries or road-cuttings, others are much smaller and temporary, such as pipeline trenches and excavations for new buildings. A number of these sites stand apart from the rest by the variety, abundance, rarity or fine preservation of the fossils they have yielded. Some of these 'classic localities' are still accessible, but others have long since disappeared.

The nine featured in the following pages are chosen to span the geological column from Ordovician to Palaeogene, and the country from north to south. The fossils illustrated are from the collections of the British Geological Survey unless otherwise stated, and are reproduced at half their natural size.

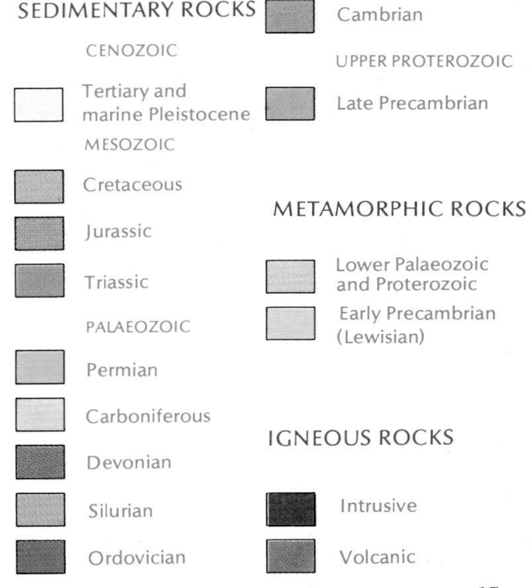

Geological map of Britain and Ireland
The map shows sedimentary rocks classified according to their age of deposition and igneous rocks according to their mode of origin.

SEDIMENTARY ROCKS

CENOZOIC
- Tertiary and marine Pleistocene

MESOZOIC
- Cretaceous
- Jurassic
- Triassic

PALAEOZOIC
- Permian
- Carboniferous
- Devonian
- Silurian
- Ordovician
- Cambrian

UPPER PROTEROZOIC
- Late Precambrian

METAMORPHIC ROCKS
- Lower Palaeozoic and Proterozoic
- Early Precambrian (Lewisian)

IGNEOUS ROCKS
- Intrusive
- Volcanic

TONY STONE

trilobite : *Ogygiocarella debuchii* (×½)

brachiopod
Palaeoglossa attenuata (×¾)

A KEENAN

trilobite with large eye
Microparia lusca (×3)

trilobite head shields
Trinucleus fimbriatus (×1)

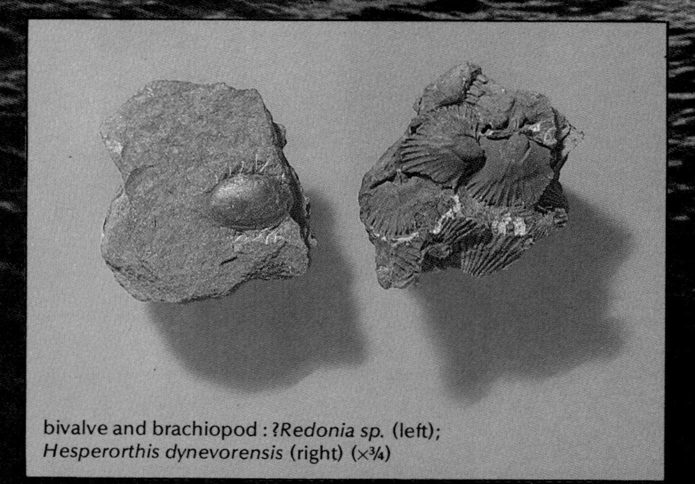
bivalve and brachiopod : ?*Redonia sp.* (left); *Hesperorthis dynevorensis* (right) (×¾)

background : seascape

BUILTH

Rugged country northeast of Builth Wells is made up of volcanic and sedimentary rocks of Ordovician age, well exposed in crags, quarries and river sections. There are 3000m of strata present in the area, forming an 'inlier' of old rocks surrounded by younger formations. The rocks and their fossils have been studied in great detail, revealing a complex story of changing environments during the period from about 450 to 470 million years ago. An Ordovician shoreline, complete with cliffs, sea stacks and beach boulders has even been identified.

Three types of rocks from different marine environments present in the area at different times can be recognised by the fossils: deep-water shales containing graptolites; medium-depth silty shales containing graptolites, trilobites and occasional gastropods and brachiopods; and shallow-water sandstones containing brachiopods and sponges. The beach beds are virtually unfossiliferous.

Other famous localities for Ordovician fossils are Abereiddy Bay, Dyfed, for graptolites, and Shelve, Shropshire, for trilobites and brachiopods.

sponge-anchoring spicules *Pyritonema fasciculus* (×½)

graptolite, zonal index species *Didymograptus murchisoni* (×¾)

trilobites
a *Cnemidopyge nuda*
b *Platycalymene duplicata*
c *Lloydolithus lloydi*
(×¾)

graptolite, zonal index species *Glyptograptus teretiusculus* (×3)

ocean
deep water
Shelve
Builth Wells
Abereiddy Bay
shallow water

geography of middle Ordovician times

DUDLEY

The Wren's Nest, Dudley, is a low, elongated hill made up of a core of shale surrounded by massive limestone, the Wenlock Limestone. This was laid down in the middle of the Silurian Period, about 420 million years ago, when a shallow sea extended across much of central and southern England. Corals, bryozoans and stromatolites built up reefs which must have stood out as low mounds on the sea floor; brachiopods, crinoids, trilobites and gastropods lived around them, and nautiloids swam above.

Limestone was mined and quarried at the Wren's Nest from the end of the 18th century until 1924, leaving huge underground caverns, many now in danger of collapse. The fossils in the limestone first attracted attention 200 years ago, particularly the trilobite *Calymene*, which is featured on the town's coat of arms and is commonly called the 'Dudley locust'. The site is now a National Nature Reserve, in the care of the Nature Conservancy Council.

Other famous localities for Wenlock Limestone fossils are Wenlock Edge, Shropshire, and May Hill, Gloucestershire. The fossils are illustrated half-size.

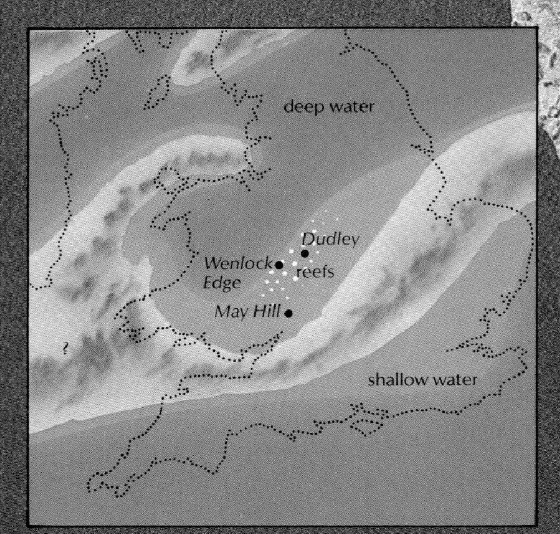

geography of middle Silurian times

colonies of tabulate corals : *Favosites sp.* (top)
Halysites catenularius, 'chain coral' (bottom)

sea snails
Poleumita discors (top)
Platyceras haliotis (bottom)

crinoid, specimen lacking roots
Dimerocrinites icosidactylus

crinoid, fine complete specimen : *'Taxocrinus' sp.*

trilobite, rolled up for protection *Calymene blumenbachii*

background : Wenlock Limestone from Dudley

trilobites : *Dalmanites sp.*, with prominent compound eyes (left); *Calymene blumenbachii*, the 'Dudley locust' (right)

brachiopods : a *Dolerorthis rustica*; b *Atrypa reticularis*; c *Gypidula galeata*; d *Leptaena depressa*

solitary rugose coral *Cystiphyllum siluriense*

COLEMAN

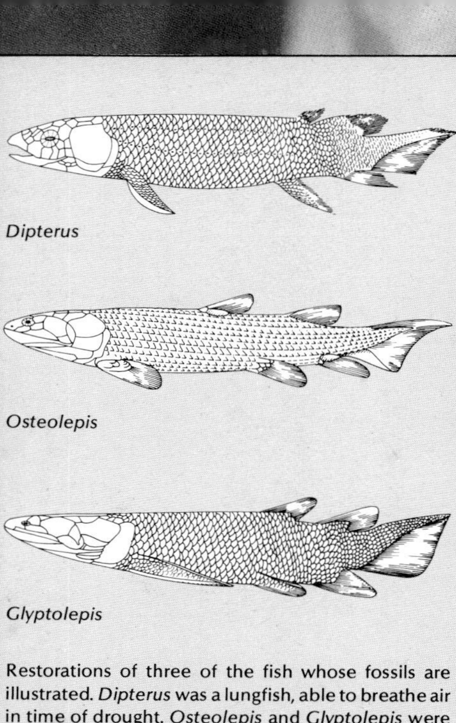

Restorations of three of the fish whose fossils are illustrated. *Dipterus* was a lungfish, able to breathe air in time of drought. *Osteolepis* and *Glyptolepis* were both bony fish of the rhipidistan group. *Thursius*, illustrated opposite, was similar to *Osteolepis* in general appearance.

skull roof
and underside
showing teeth
Dipterus valenciennesi

ROYAL SCOTTISH MUSEUM

Osteolepis panderi

Dipterus macrolepidotus

background : European freshwater fish

THURSO

The cliffs around Thurso are formed by sandstones of the Upper Caithness Flagstone Group that were laid down in the middle of the Devonian Period, about 360 million years ago. The only fossils are rare fish scales and plant fragments, except in thin bands of black calcareous flagstone where complete fossil fish are found. Fossils from these fish beds were first collected seriously by Robert Dick, a baker of the town, in the 1840s, and were made famous by Hugh Miller in his book *Footsteps of the Creator*, 1849, which he wrote to combat the idea of evolution.

The rocks of Thurso were laid down in a series of freshwater lakes which stretched from the Moray Firth to the Shetland Islands. The rest of central and northern Britain was a hot and mountainous landmass, part of the huge continent of Euramerica. The fish beds represent periods when the bottom waters of the lake became stagnant and free from scavengers and when fish were abundant in the surface waters. Water level in the lake fluctuated, and at times huge areas dried up, killing the fish. There are other classic localities at Nairn, Banff and Achanarras. The fossils are illustrated half-size.

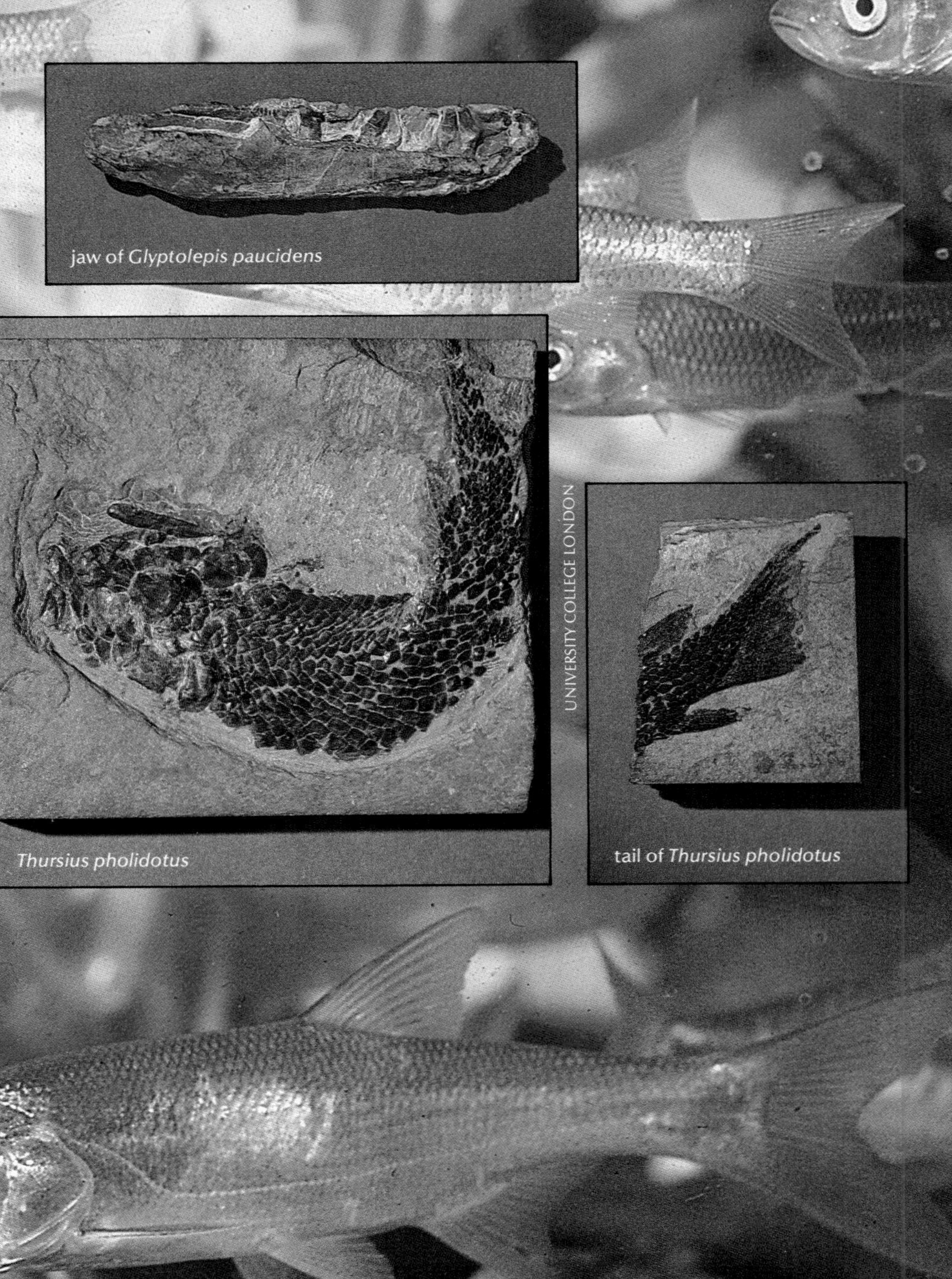

jaw of *Glyptolepis paucidens*

Thursius pholidotus

tail of *Thursius pholidotus*

UNIVERSITY COLLEGE LONDON

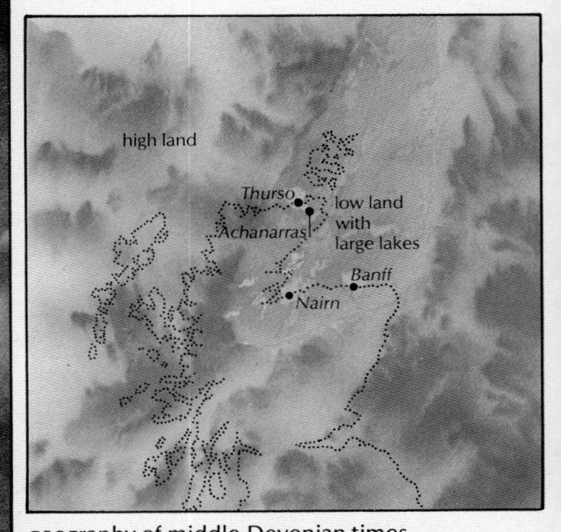

geography of middle Devonian times

BGS L196

brachiopods : a *Spirifer striatus*; b *Antiquatonia hindi*; c *Echinoconchus punctatus*

BM (NH)

orthoconic nautiloid, related to living nautilus

A E TIMMS

sea snails : a *Straparollus dionysii*; b *Naticopsis elliptica*; c *Straparollus pentangulatus*

background : Winnat's Pass, Castleton

A E TIMMS

bivalves, one showing colour-banding : a *Aviculopecten plicatus* (left); *Streblochondria ellipticum* (right)

BM (NH)

brachiopods : a *Dielasma hastata*; b *Martinia glabra*; c *Pleuropugnoides pleurodon* (×¾)

goniatite : *Goniatites striatus*

BM (NH)

trilobite
head and tail shields
Phillipsia sp (×1)

CASTLETON

Castleton is famous for its caves and for the banded fluorite called 'Blue John' that they contain. It is also famous for the early Carboniferous fossils found in the limestones that crop out at Treak Cliff, Cow Low, Cave Dale and other localities around the village. The limestones, of Viséan age, were laid down under the sea 330 million years ago. South of Castleton the sea water was shallow and calm, inhabited by crinoids, corals and some brachiopods. To the north the sea was at least 150m deep, and supported a fauna of goniatites, gastropods, bivalves, trilobites and bryozoans. At Castleton itself a long narrow reef, built up nearly to the water surface by algae, and inhabited by a great variety of animals, must have formed an effective barrier between the basin and shelf. The reef did not persist unchanged for long, but died and was re-established twice as the water level changed.

Other famous localities for Carboniferous Limestone fossils include Clifton and the Avon Gorge, Bristol, and Clitheroe, Lancashire. The fossils are illustrated half-size except where otherwise indicated.

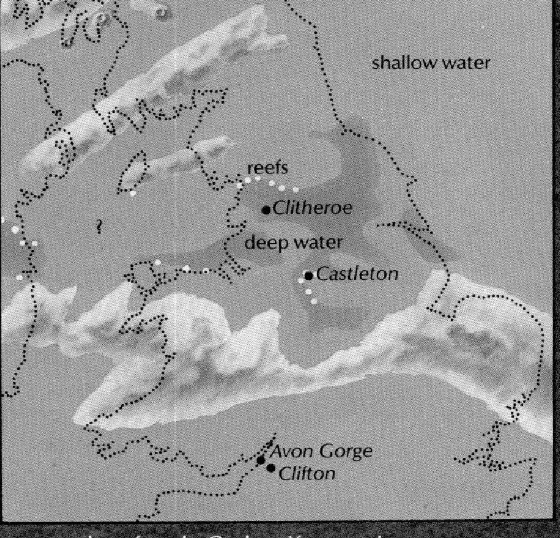

geography of early Carboniferous times

BARNSLEY

Of the many collieries around Barnsley, the most famous for its fossils was Monckton Main, just north of the town. Its shafts penetrate 600m of sandstones, mudstones, fossil soils and coal seams of Westphalian (late Carboniferous) age, which were laid down around 300 million years ago. Fossil plants are common in the mudstones and in addition there are beds with freshwater shells (mussel bands) or marine bivalves and goniatites (marine bands). The specimens illustrated were collected from the mudstones around the Barnsley Coal by W. Hemingway, who lived in the town at the end of the last century.

300 million years ago Barnsley was situated on a low-lying plain on which river channels meandered through swamps and lagoons. Slight lowering of sea level led to the emergence of wide mud flats, which in time became covered by a forest of ferns, seed ferns, giant clubmosses and horsetails. A rise in sea level led to flooding of the plain and the deposition of a thin layer of marine sediment.

Other famous localities for Carboniferous plants are Radstock, Avon, and Coseley, West Midlands. The fossils are illustrated half-size.

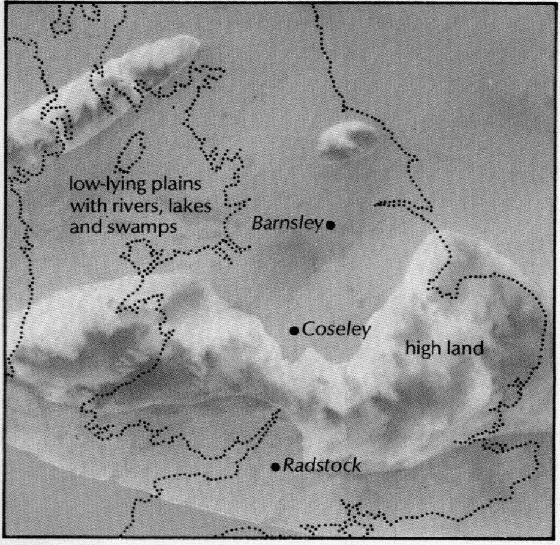

geography of late Carboniferous times

leafy twig of a clubmoss
Bothrodendron minutifolium

base of horsetail stem
Calamites suckowi

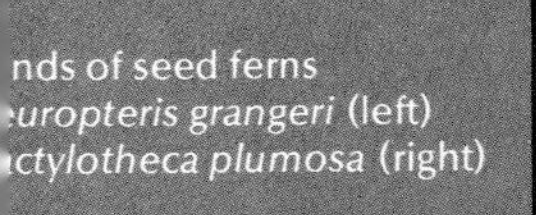

nds of seed ferns
europteris grangeri (left)
ctylotheca plumosa (right)

frond of a fern
Pecopteris bioti

frond of a seed fern : *Alethopteris lonchitica*

leaf scars on club moss trunk
Sigillaria rugosa

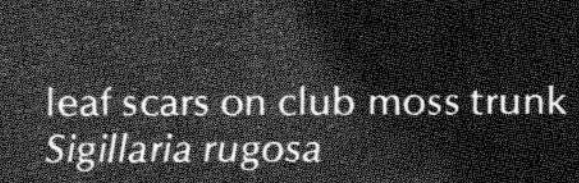

cone of a club moss
Sigillariostrobus rhombibracteatus

background : forest swamp, South America

COLEMAN

LYME REGIS

Cliffs around Lyme Regis display a succession of highly fossiliferous early Jurassic rocks called the Lower Lias. These are black and grey shales, with alternating bands of limestone and shale (Blue Lias) in the lower part of the succession. They were laid down 190 million years ago under a sea which was progressively submerging much of England and Wales. The most common fossils are the remains of swimming and floating animals such as ammonites, belemnites, thin-shelled bivalves, fish and marine reptiles (ichthyosaurs and plesiosaurs). Remains of burrowing animals and those that lived on the sea floor are uncommon at some horizons, suggesting that at times the bottom waters were stagnant and unfit for life. Logs of wood that drifted out to sea from a nearby landmass seem to have provided points of attachment for such animals as crinoids, brachiopods and bivalves that normally lived on the sea floor.

Lyme Regis has been famous for its fossils at least since 1811, when Mary Anning, then aged 12, excavated a nearly complete ichthyosaur skeleton from the cliffs. Other famous localities for Lower Lias fossils are Cheltenham and Robin Hood's Bay. The fossils are illustrated half-size.

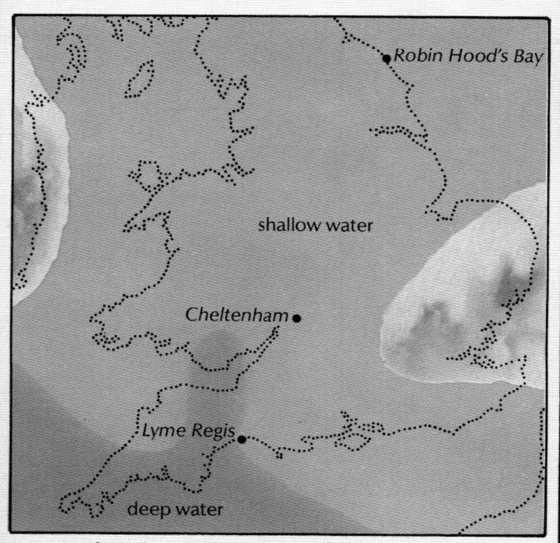

geography of early Jurassic times

sea snail and bivalves
a *Amberleya acuminata*
b *Gryphaea arcuata*
c *Ctenostreon tuberculatus*

bed of crinoids : *Pentacrinites fossilis*

bed of ammonites
Promicroceras planicosta

bony fish : *Pholidophorus bechei*

belemnites
guards of extinct squids

ammonites : a *Oistoceras wrighti*; b *Asteroceras obtusum* (zonal index species); c *Tragophylloceras loscombi*

background : cliffs east of Lyme Regis

MALTON

The Howardian Hills, to the west and south of Malton, are composed of late Jurassic limestones and sandstones, known as the Corallian. The lowest part of the 55m succession consists of calcareous sandstone, rich in sponge spicules but not otherwise very fossiliferous; this formation is the same age as the uppermost Oxford Clay of southern England. Above this is the Coralline Oolite Formation, which includes fossiliferous sandstone, limestone rich in large bivalves, gastropods, and sea urchins, the remains of coral reefs, and oolite (limestone made up of tiny spherical grains) with occasional finely preserved fossils. At the top of the succession is a calcareous shale.

Fossils from this area were easy to collect during the last century when many small stone quarries were in use, but today Corallian fossils are more accessible on the coast from Scarborough to Filey Brigg. Another famous locality for these fossils is the cliffs near Weymouth, Dorset. The fossils are illustrated half-size.

sea urchins
a spine of *Plegiocidaris sp.*
b *Pseudodiadema pseudodiadema*
c *Nucleolites scutatus*

sea snail : *Bourguetia saemanni*

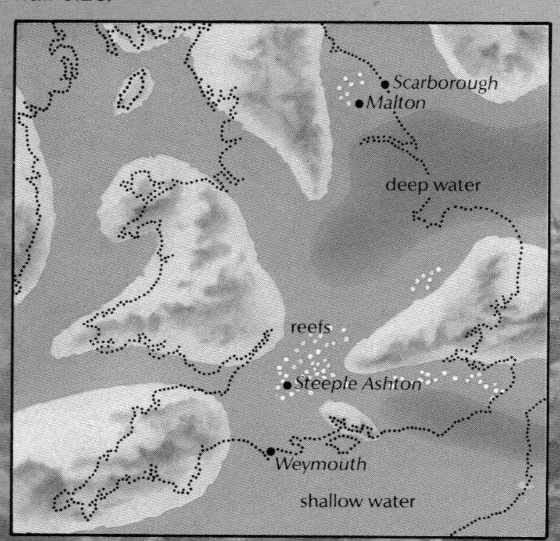

geography of late Jurassic times

belemnite, guard of an extinct squid *Pachyteuthis sp.*

bivalves
a *Chlamys fibrosa*
b *Chlamys splendens*
c *Perampliata ampliata*
d *Mytilus ungulatus*
e *Trigonia reticulata*

corals : *Montlivaltia* cf *melania*

rhynchonellid brachiopod *Thurmannella thurmanni*

sea snail : *Nerinaea roemeri*

background : coral reef, Indian Ocean

COLEMAN

starfish : *Calliderma smithiae* Lower Chalk

ammonites
a *Euhoplites opalinus*; b *Heteroclinus nodosus*; c *Anahoplites planus*; d *Hoplites dentatus*; e *Hamites maximus* Gault Clay

sea snails : a *Anchura carinata*; b *Jurassiphorus fittoni*;

bivalve : c *Nucula pectinata* Gault Clay

sponge : *Exanthesis labrosus* Lower Chalk

sea urchins : *Holaster subglobosus* (left); *Hemiaster morrisii* (right) Lower Chalk

worm tube : *Serpula umbonata* (left)
brachiopod : *Concinnithyris subundata* (righ
Lower Chalk

FOLKESTONE

Two fossiliferous formations are accessible east of Folkestone: the Lower Chalk, which forms the cliffs towards Dover, and the underlying Gault Clay, which is visible around Folkestone itself and in the huge landslip just east of the town. Fossils are beautifully preserved in the fresh Gault Clay and have a pearly lustre, although fresh clay has become hard to find since the sea defences were built in the 1950s. In the Lower Chalk, on the other hand, shells made of aragonite have dissolved, leaving ammonites, gastropods and some bivalves preserved as casts. These rocks were laid down in the middle of the Cretaceous Period. The Gault Clay sea was warm, between 150 and 200m deep, rather muddy and with a stiff clay floor. Most of the animals were small and thin-shelled. The Lower Chalk sea was also warm, but clear, rather deeper than the Gault sea and with a very soft floor. Sea urchins and crustaceans burrowed into the soft lime mud; sponges, starfish, brachiopods and bivalves lived on the sea bed, and ammonites floated above. The fossils are illustrated half-size.

bivalves : *'Pecten' beaveri* (left); *Camptonectes dubrisiensis* with worm tubes and an oyster shell (right) Lower Chalk

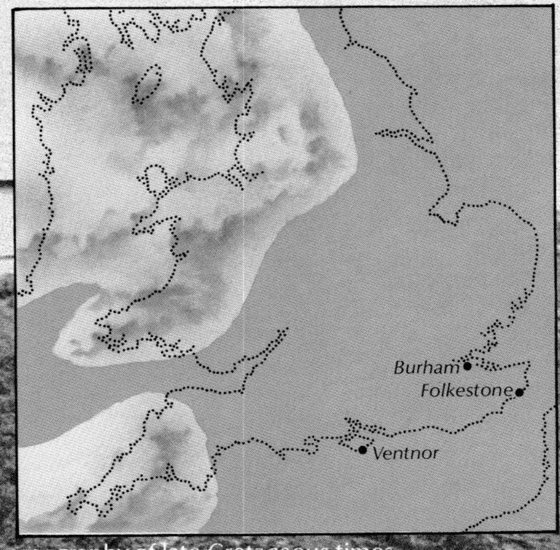

geography of late Cretaceous times

ackground : Folkestone Warren

BARTON

Low cliffs on either side of the village of Barton are made up of silty and sandy clays of late Eocene age called the Barton Beds. Layers in the cliff face slope gently down to the east, so the oldest beds are west of the village. Fossils are generally plentiful, though they are crushed or very fragile at some horizons. Where they are well preserved, complete fossils can be extracted with a small trowel or broad-bladed knife. The Barton Beds were laid down only about 40 million years ago in a sea which was probably not more than 50m deep and which extended from France into southern England. The most common animals were bivalves, most of which burrowed in the mud, and gastropods, which grazed or preyed on bivalves and worms; sharks and rays swam in the water above. Towards the end of the period during which the beds were deposited, the water became shallower and brackish, inhabited only by the few molluscs that tolerated such an environment.

Other good exposures of Barton Beds are seen in Alum and Whitecliff bays on the Isle of Wight. The fossils are illustrated half-size.

sea snails : a *Sycostoma pyrus*; b *Fusinus porrectus*; c *Hastator imbricatarius*; d *Cornulina minax*; e *Sconsia ambigua*

sea snail : *Athleta luctator*

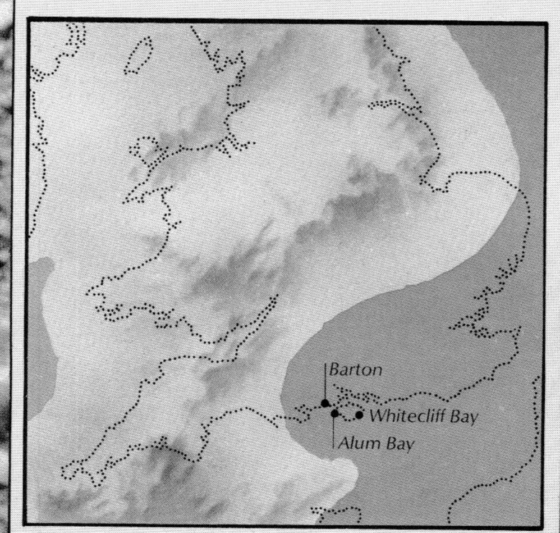

geography of middle Palaeogene times

COLEMAN

bivalves
a *Azor compressa*
b *Crassatella sulcata*
c *Glycymeris deleta*
d *Chama squamosa*

sharks' teeth
a *Jaekelotodus trigonalis*
b *Synodontaspis hopei*

scaphopod
Dentalium bartonense

sea snail
Clavilithes macrospira

sea snail
Hippochrenes amplus

sea snail
Ampullonatica ambulacrum

background : shells on a sandy beach

Foreign fossils

Fossiliferous rocks of all the twelve geological systems are found in Britain, although the record is incomplete. Some gaps exist because strata were laid down and have since been buried or eroded away, some because there was no sedimentation over Britain at the time. One of the gaps is at the junction of the Palaeogene and Neogene systems. The youngest rocks in the Hampshire Basin are the Hamstead Beds, which are middle Oligocene; the oldest beds in East Anglia are the late Pliocene Coralline Crag. Only a few tiny pockets of sediment in Kent and some fossils washed into a Pleistocene conglomerate in Suffolk have survived from this 30-million-year period. Thick fossiliferous sequences dating from this period are however to be found in many other parts of the world, including Belgium and northern France, Italy, Malta and parts of the USA (fig 33). There is a similar gap in the British succession at the junction of the Cretaceous and Palaeogene systems.

Even where rocks of a particular age are accessible in Britain they seldom represent the full range of environments and therefore do not contain all the fossils of that age that are known. Study of foreign fossils is necessary to complete the picture. Early and middle Triassic rocks in Britain were laid down on land or in fresh or very salty water and contain fossil reptiles, fish, scorpions, tiny water fleas and plants. Marine rocks containing brachiopods, a few corals, bivalves, sea snails and animals very like ammonites are found in Switzerland, Austria, Yugoslavia (fig 34) and elsewhere. Similarly, the Chalk of southern England was laid down under the sea and is rich in fossil molluscs, echinoids and fish; only very rarely did the carcass of a land animal drift into the area. Rocks of the same age in western North America were laid down in river deltas and flood plains and contain fossil dinosaur bones, plant remains and freshwater shells and fish.

Many groups of fossils are not found in Britain even though fossiliferous rocks laid down in a suitable environment are accessible. Their absence may be due to ancient physical barriers or climatic factors. Separate 'faunal realms' develop whenever a large tract of land or sea becomes isolated by a mountain range, desert or an ocean deep, and the organisms within it start to evolve differently from those outside. Redlichiid trilobites were abundant in early Cambrian times, but were confined to a realm that included Asia, Australia and Antarctica. They are not found in Britain, which at that time was part of an American-Scandinavian faunal realm in which olenellid trilobites flourished. Archaeocyathids (fig 35), early Cambrian sponge-like animals that formed large reefs, lived in both these realms, but are still not found in Britain. The ancient climate may have been the cause of this: it is possible that the water over Britain was too cold for them to flourish.

BGS

33 Miocene gastropod from Maryland, USA

BGS

34 Triassic ammonoid from Yugoslavia

BM(NH)/BGS

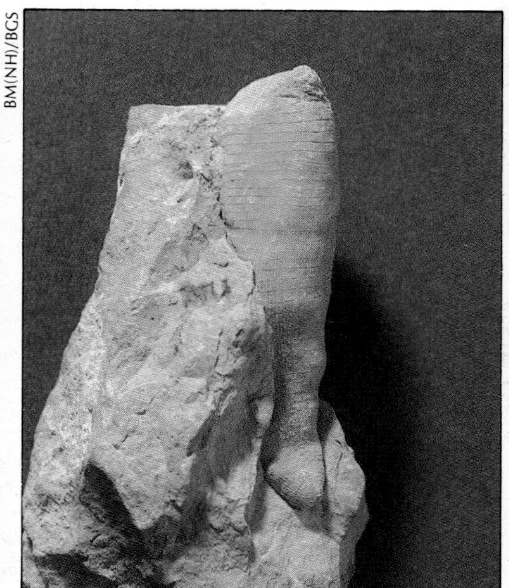

35 Cambrian archaeocyathid from Morocco